Refah Saad Alkhaldi

Calculating the Position of Electrons Around the Nucleus

Refah Saad Alkhaldi

Calculating the Position of Electrons Around the Nucleus

Demonstrating the Orbital Shape of Hydrogen Atom and Helium Atom

LAP LAMBERT Academic Publishing

Imprint
Any brand names and product names mentioned in this book are subject to trademark, brand or patent protection and are trademarks or registered trademarks of their respective holders. The use of brand names, product names, common names, trade names, product descriptions etc. even without a particular marking in this work is in no way to be construed to mean that such names may be regarded as unrestricted in respect of trademark and brand protection legislation and could thus be used by anyone.

Cover image: www.ingimage.com

Publisher:
LAP LAMBERT Academic Publishing
is a trademark of
International Book Market Service Ltd., member of OmniScriptum Publishing Group
17 Meldrum Street, Beau Bassin 71504, Mauritius
Printed at: see last page
ISBN: 978-620-3-19825-6

Calculating the Position Of Electrons Around the Nucleus and Demonstrating the Orbital Shape of Hydrogen Atom and Helium Atom

By: Refah Saad Alkhaldi

rkhaldi@psmchs.edu.sa

<u>Contents:</u>

Atomic Theory:

It is the scientific theory that matter is composed of particle called atoms.[1]

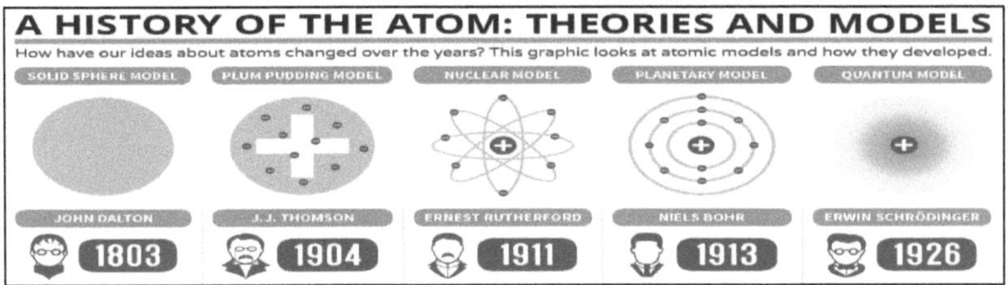

Figure:1

John Dalton:[1]

In the early 1800s, the scientist John Dalton noticed the chemical substances seemed to combine and break down into other substances by weight in proportions that suggested that each chemical element is ultimately made up of tiny indivisible particles of consistent weight.

Figure: 2   Figure:3

Shortly after 1850s, certain physists developed the kinetic theory of gases and of heat, which mathematically modeled the behavior of gases by assuming that they were made of particles.

Near the end of the 18th century two laws about chemical reactions emerged without referring to the notion of an atomic theory.

First, was the law of conservation of mass, closely associated with the work of Antoine Lavoisier, which states that the total mass in a chemical reaction remains constant (that is, the reactants have the same mass of the products).

Figure: 4

Second, was the law of definite proportions. First established by the French chemist Joseph Proust in 1797, this law states that if a compound is broken down into its constituent chemical elements, then the masses of the constituents will always have the same proportions by weight, regardless of the quantity or source of the original substance.

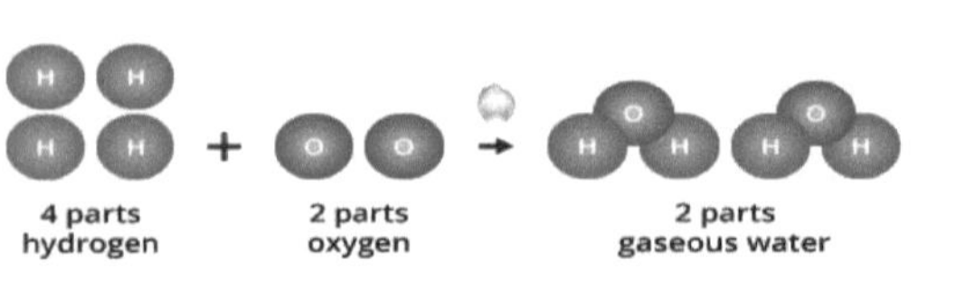

Figure:5   Figure: 6

John Dalton studied and expanded upon this previous work and defended a new idea, later known as the law of multiple proportions: if the same two elements can be combined to form number of different compounds, then the ratios of the masses of the two elements in their various compounds will be represented by small whole numbers. This is a common pattern in chemical reactions that was observed by Dalton and other chemists at the time.

Philosophical Atomism:[1]

The idea that the matter is made up of discrete units is a very old idea, appearing in many ancient cultures such as Greece and India.

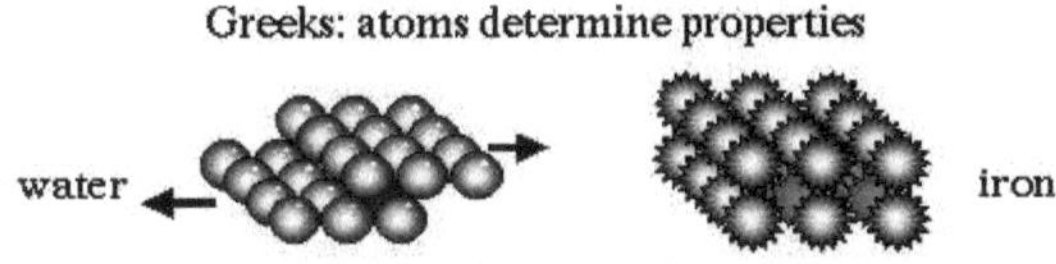

Figure :7

The word "atom", meaning uncuttable, was coined by the Pre-Socratic Greek philosophers Leucippus and his pupil Democritus [c.460–c.370 BC].

True knowledge exists in knowing that you know nothing.

Socrates

Figure:8

Democritus thought that atoms were infinite in number, uncreated, and eternal, and that the qualities of an object result from the kind of atoms that compose it.

Figure:9

Figure:10

Democritus's atomism was refined and elaborated by the later Greek philosopher Epicurus [341-2070 BC], and the Roman Epicurean Poet Lucretius [c. 99- c.55 BC].

Figure: 11

Figure:12

During the Early Middle Ages, atomism was mostly forgotten in western Europe. During the 12th century, atomism became known again in western Europe through references to it in the newly rediscovered writings of Aristotle.

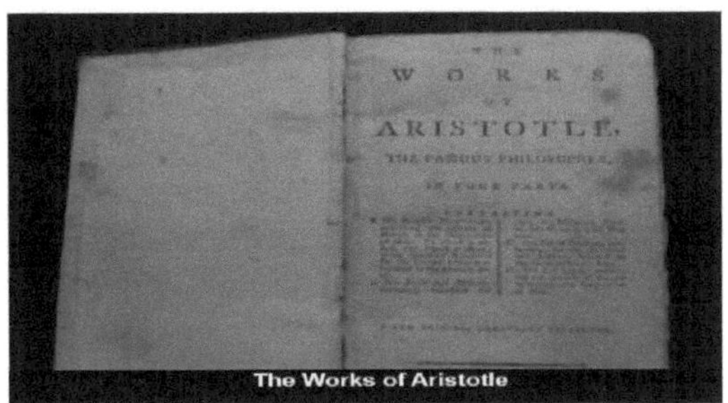

Figure:13

In the 14th century, the discovery of major works describing atomist teachings, including Lucretius's De rerum natura and Diogenes Laertius's Lives and opinions of Eminent Philosophers, led to increased scholarly attention on the subject.

Figure:14

Nonetheless, because atomism was associated with the philosophy of Epicureanism which contradicted orthodox Christian teachings, belief in atoms was not considered acceptable by most European philosophers.

The French Catholic Priest Pierre Gassendi; (1592 – 1655) revived Epicurean atomism with modifications, arguing that atoms were created by God and, through extremely numerous, are not infinite and the first person who use term "molecule" to describe aggregate of atom. Gassendi's modified theory of atoms was popularized in France by the Physician Francois Bernier (1620 – 1688) and in England by the natural philosopher Walter Carleton (1619 – 1707).

Figure:15

The Chemist Robert Boyle (1627 – 1691) and the Physists Isaac Newton (1642 – 1727) both defined atomism, and by the end of the 17th century, it had become accepted by points of the scientific community.

Figure: 16

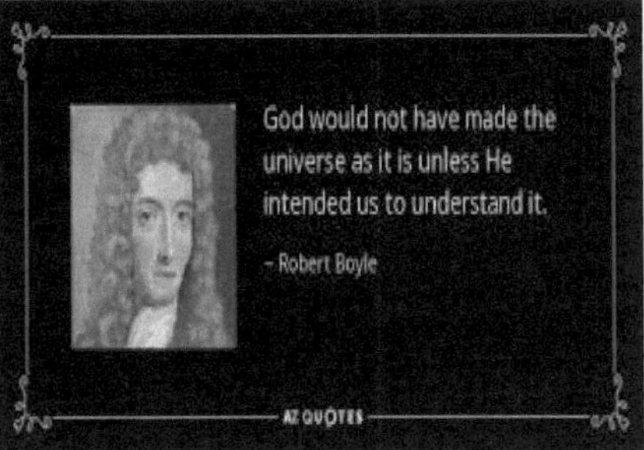

Figure: 17

In the early 20th century, Albert Einstein and Jean Perrin Proved that the Brownian motion (the erratic motion of Pollen grains in water) is caused by the action of water molecules; this third line of evidence silenced remaining doubts among scientists as to whether atoms and molecules were real.

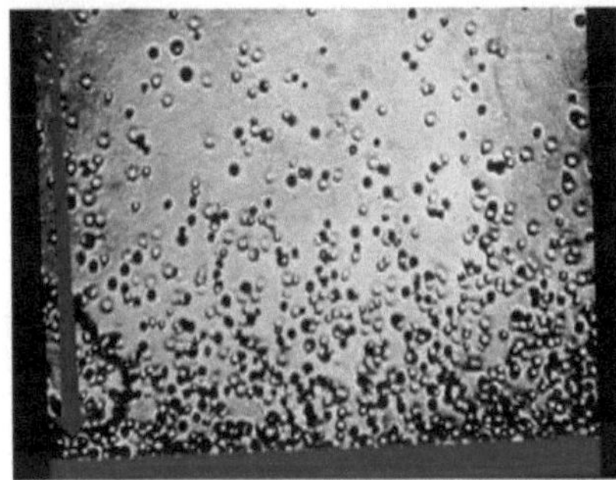

A microphotography of the height distribution of particles of resin suspended in water. This photograph was taken by Jean Perrin and subsequently exhibited in the science museum he founded in Paris in 1937, the Palais de la Découverte (© Palais de la Découverte, Paris).

Figure:18   Figure:19

Throughout the 19th century, some scientists had cautioned that the evidence for atoms was indirect, and therefore atoms might not actually be real, but only seem to be real.

By the early 20th century, scientists had developed fairly detailed and precise models for the structure of matter, which led to more rigorously defined classifications for the tiny invisible particles that make up ordinary matter.

An atom is now defined as the basic particle that composes a chemical element. Around the turn of 20th century, physicists discovered that the particles that chemists called "atom" are in fact agglomerations of even smaller particles (subatomic particles), but scientists kept the name out of convention. The term elementary particle is now used to refer to particles that are actually invisible.

Thomson Atomic Model:[2]

Earliest theoretical description of the inner structure of atoms, proposed about 1900 by William Thomson (Lord Kelvin) and strongly supported by Sir Joseph John Thomson, who had discovered (1897) the electron, a negatively charged part of every atom. Thomson held that atoms are uniform spheres of positively charged matter in which electrons are embedded. Popularly known as the Plum Pudding Model.

Figure:20

Figure:21

Rutherford Model:[3]

Thomson Plum Pudding model was disproved in 1909 by one of his former students, Ernest Rutherford, who discovered that the most of the mass and positive charge of an atom is concentrated in a very small fraction of its volume, which he assumed to be at very center.

Figure: 22

Earnest Rutherford and his colleagues Hans Geiger and Ernest Marsden came to have doubts about the Thomson model after they encountered difficulties when they tried to build on instrument to measure the charge-to-mass ratio of alpha particles (these are positively-charged particles emitted by certain radioactive substances such as radium).

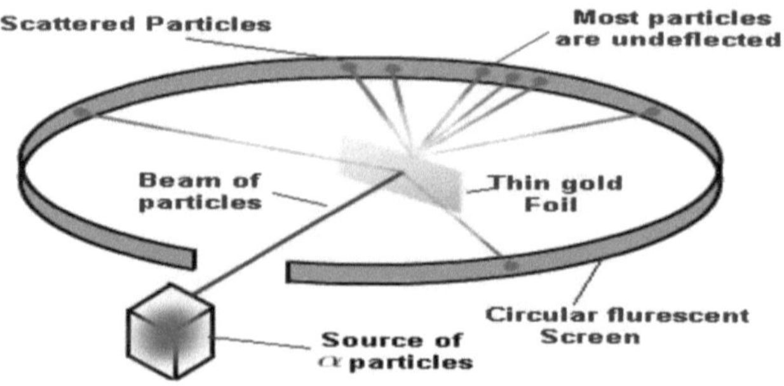

Figure: 23

The alpha particles were being scattered by the air in the detection chamber, which made the measurement unreliable. Thomson has encountered a similar problem in his work on cathode rays, which solved by creating a near perfect vacuum in his instruments. Rutherford didn't think he'd run into this same problem because alpha particles are much heavier than electrons.

Between 1908 and 1913, Rutherford and his colleagues performed a series of experiments in which they bombarded thin foils of metal with alpha particles. They spotted alpha particles being deflected by angles greater than 90 degree.

To explain this, Rutherford proposed that the positive charge of the atom is not distributed throughout the atom's volume as Thomson believed, but is concentrated in a tiny nucleus at the center. Only such an intense concentration of charge cloud produces an electric field strong enough to deflect the alpha particles as observed.

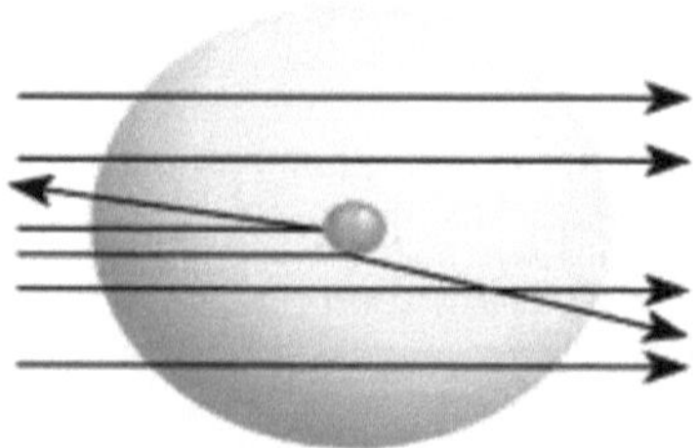

Figure: 24

Bohr Model:[4]

The planetary model of atom had two significant shortcomings:

Figure: 25

1. The first is that, unlike planets orbiting a sun, electrons are charged particles. An accelerating electric charge is known to emit electromagnetic waves according to the Larmor formula in classical electromagnetism. An orbiting charge should steadily lose energy and spiral toward the nucleus, colliding with it in a small fraction of a second.
2. The second problem was that the planetary model could not explain the highly peaked emission and abruption spectra of atoms that were observed.

In 1913, Niels Bohr incorporated this idea into his Bohr model of the atom, in which an electron could only orbit the nucleus in particular circular orbits with fixed angular momentum and energy, its distance from the nucleus (i.e. their radii) being proportional to its energy.
Under this model an electron could not spiral into the nucleus because it could not lose energy in a continuous manner; instead, it could only make instantaneous "quantum leaps" between the fixed energy levels. When this occurred, light was emitted or absorbed at a frequency proportional to the change in energy; hence the absorption and emission of light in discrete spectra).

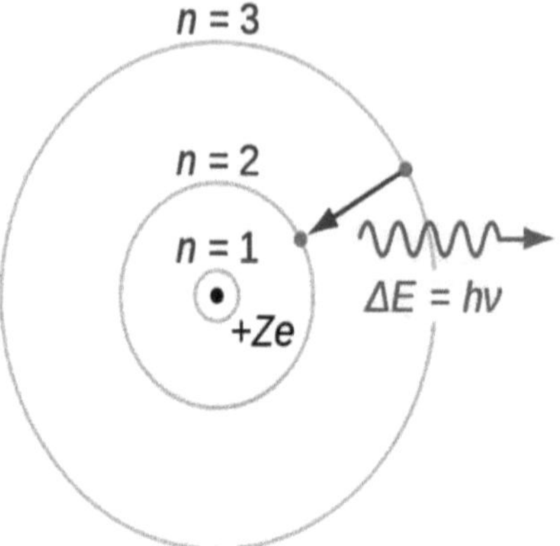

Figure: 26

Bohr's model was not perfect. It could only predict the spectral lines of hydrogen; it could not predict those of multielectron atoms.
Worse still, as spectrographic technology improved, additional spectral lines in hydrogen were observed which Bohr's model couldn't explain.

Figure: 27

In 1916, Arnold Sommerfeld added elliptical orbits to the Bohr model to explain these extra emission lines, but this made the model very difficult to use, and it still couldn't explain more complex atoms.

Quantum theory revolutionized physics at the beginning of the 20th century, when Max Planck and Albert Einstein postulated that the light energy is emitted or absorbed in discrete amounts known as quanta (singular, quantum).

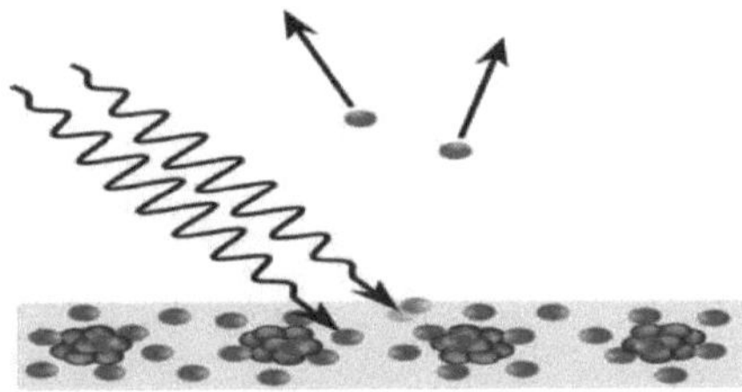

Figure:28

Erwin Schrödinger:[5]

Schrödinger began to think about explaining the movement of an electron in an atom as a wave. By 1926 he published his work, providing a theoretical basis for the atomic model that Niels Bohr had proposed based on laboratory evidence.

Figure:29

The equation at the heart of his publication became known Schrödinger's wave equation. Thus, the second theoretical explanation of electrons in an atom, following Werner Heisenberg's matrix mechanics.

$$\frac{-\hbar^2}{2m}\nabla^2\Psi(r) + V(r)\Psi(r) = E\Psi(r)$$

$$\underset{\substack{Kinetic\\Energy}}{} + \underset{\substack{Potential\\Energy}}{} = \underset{\substack{Total\\Energy}}{}$$

Figure: 30

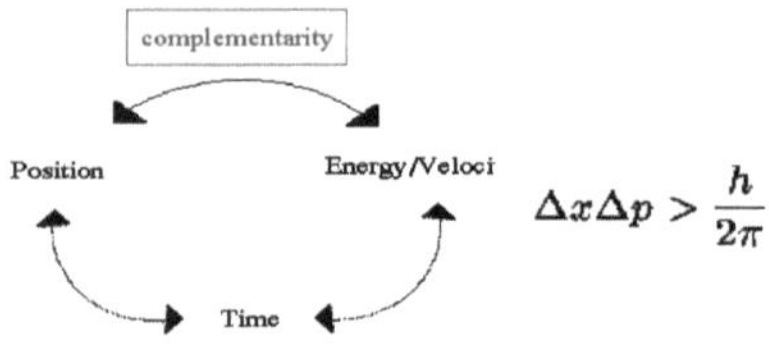

$$\Delta x \Delta p > \frac{h}{2\pi}$$

Figure: 32   Figure: 33

Many scientists preferred Schrödinger theory since it could be visualized, while Heisenberg's was strictly mathematical. Both are identical theories, but only expressed differently.

Orbital Shapes:[6]

The hydrogen atom orbitals are the "wavefunction" portion of the quantum mechanical solution to the hydrogen atom. The wave functions tells us about the probability of finding the electron at a certain point in space. Thus, the orbitals offer us a picture of the electron in a hydrogen atom. However, this picture is not a simple one, for the following reasons:

- First, the electron is spread out over space. The one cannot pin-point its location. If this seems odd (even impossible), it should.
- Second, Electrons are inherently quantum mechanical and behave differently than every thing in our everyday life .

As the one builds up from hydrogen atom to multi-electron atom to bonding and molecules, the one will take extensive use of the wavefunctions that the first find for hydrogen. As such it is useful to become familiar with their shapes.

There are two key features for an orbital:
- The distribution of the electron away from the nucleus. This is known as radial distribution.
- The other is the "shape" of the orbital. And this is known as the angular distribution.

The radial distribution is mostly dependent on the principle quantum number "n". While the angular distribution depends on "l" and "m_l".

Atomic Orbitals:[7]
Orbitals and orbits:

The simple view of the electron is orbiting around the nucleus in an atom looks similar to the planet moves around the sun. But the truth is different and electrons in fact in habit of space known as orbitals.

The impossibility of drawing orbits for electrons: to plot a path for something you need to know exactly where the object is and be able to work out where exactly where it is going to be an instant later.

YOU CANNOT DO THIS FOR ELECTRONS.

The Heisenberg Uncertainty principle says –loosely – that you cannot know with certainty both where an electron is and where it is going next. (What actually says that it is impossible to define with absolute precision, at the same time, both the position and the momentum of an electron).

That makes it impossible to plot an orbit for an electron around a nucleus.

Uncertainty Principle:[8]

Orbital Shapes In quantum mechanics:

In quantum mechanics the uncertainty principle (also known as Heisenberg's Uncertainty Principle) in any of a variety of mathematical inequalities asserting a fundamental limit to the accuracy with which the values for certain pairs of physical quantities of a particle, can be predicted from initial conditions.

The uncertainty principle implies that it is in general not possible to predict the value of a quantity with arbitrary certainty, even if all initial conditions are specified.

Introduced first in 1927 by German Physicist Werner Heisenberg, the Uncertainty Principle states that the more precisely position of some particle is determined, the less precisely its momentum can be predicted from initial conditions, and vice versa.

The uncertainty principle actually states a fundamental property of quantum systems and is not a statement about the observational success of current technology. It must be emphasized that the measurement does not mean only a process in which a physicist observe takes part, but rather any interaction between classical and quantum objects regardless of any observer.

Calculating The Position Of Electrons Around The Nucleus And Demonstrating The Orbital Shape Of Hydrogen Atom And Helium Atom

By: Refah Saad Alkhaldi

Introduction:[9]

In atomic theory and quantum mechanics, an atomic orbital is a mathematical function describing the location and wave-like behavior of an electron in an atom. This function can be used to calculate the probability of finding any electron of an atom in any specific region around the atom's nucleus. The term atomic orbital may also refer to the physical region or space where the electron can be calculated to be present, as predicted by a particular mathematical from the orbital. Atomic orbitals are the basic building blocks of the atomic orbital model (alternatively known as electron cloud or wave mechanics model), a modern framework for visualizing the submicroscopic behavior of electrons in matter. In this model the electron cloud of a multi-electron atom may be seen as being built up (in approximation) in an electron configuration that is a product of simpler hydrogen- like atomic orbitals.

With the development of quantum mechanics and experimental findings (such as the two slit diffraction of electrons), it was found that the orbiting electrons around a nucleus could not be fully described as particles, but needed to be explained by the wave particle duality. In this sense, the electrons have the following properties :

A. Wave-like properties:
1) The electrons do not orbit the nucleus in the manner of a planet orbiting the sun, but instead exist as standing waves. Thus, the lowest possible energy an electron can take is similar to the fundamental frequency of a wave on a string. Higher energy states are similar to harmonics of the fundamental frequency.
2) The electrons are never in a single point location, although the probability of interacting with the electron at a single point can be found from the wave function of the electron. The charge on

the electron acts like it is smeared proportional at any point to the squared magnitude of the electron's wave function.

B. Particle-like properties:

    1) The number of electrons orbiting the nucleus can only be an integer.

    2) Electrons jump between orbitals like particles.

The electrons retain particle-like properties such as; each wave state has the same electrical charge as its electron particle. Each wave state has a single discrete spin (spin up or spin down) depending on its superposition. Thus, electrons Cannot be described simply as solid particles.

An analogy might be that of a large and often oddly shaped "atmosphere" (the electron), distributed around a relatively tiny planet (the atomic nucleus). Atomic orbitals exactly describe the shape of this "atmosphere" only when a single electron is present in an atom.

When more electrons are added to a single atom, the additional electrons tend to more evenly fill in a volume of space around the nucleus so that the resulting collection (sometimes termed the atom's "electron cloud") tends toward a generally spherical zone of probability describing the electrons location, because of the uncertainty principle.

Types of orbitals:[9]

Atomic orbitals can be the hydrogen-like "orbitals" which are exact solutions to the SchrÖdinger equation for a hydrogen-like "atom" (i.e., an atom with one electron). Alternatively, atomic orbitals refer to functions that depend on the coordinates of one electron (i.e., orbitals) but are used as starting points for approximating wave functions that depend on the simultaneous coordinates of all electrons in an atom or molecule. The coordinate systems chosen for atomic orbitals are usually spherical coordinates (r, θ, z) in atoms and Cartesians (x, y, z) in a polyatomic molecules.

Method:

At θ_i and $|v_i|$ there is two components of velocity; $v_{x,i}$ and $v_{y,i}$, and we can find them as the following:

$$v_{x,i} = |v_i| \cos \theta_i$$

$$v_{y,i} = |v_i| \sin \theta_i$$

And the time t_i at each (x_i, y_i) point can be given from:

$$\tan \theta_i = \frac{y_i}{x_i}$$

$$y_i = x_i \tan \theta_i$$

$$y_i = \frac{\sin \theta_i}{\cos \theta_i} . x_i$$

$$y_i \cos \theta_i = x_i \sin \theta_i$$

And the distance is: $(y_i = v_{y,i} . t_i)$ and $(x_i = v_{x,i} . t_i)$

So, $t_i = \dfrac{x_i \sin \theta_i}{v_{y,i} \cos \theta_i}$ or $t_i = \dfrac{y_i \cos \theta_i}{v_{x,i} \sin \theta_i}$

From above if we know the velocity of electron and the angle we can be certain about the location of electron at any instant. Also, this equation helps us to find the electron since we know the whole orbit diagram.

In other words $t_i = \dfrac{x_i \tan \theta_i}{v_{y,i}}$ or $t_i = \dfrac{y_i}{v_{x,i} \tan \theta_i}$

Now, we have to convert from the Cartesian coordinate system (x_i, y_i) to the Cylindrial coordinate system (r_i, θ_i). Notice we ignore z component in the Cartesian coordinate system and in the Cylindrecal coordinate system.

$$r_i^2 = x_i^2 + y_i^2$$

And since $y_i = x_i \tan \theta_i$, so:

$$r_i^2 = x_i^2 + x_i^2 \tan^2(\theta_i)$$

Now, we can find values of x_i and y_i from the following relations:

$$x_i = \sqrt{\frac{r_i^2}{1+\tan^2(\theta_i)}} \quad \text{and} \quad y_i = x_i \tan \theta_i$$

The radius of atom's orbit is from the following: $r_i = \dfrac{n^2 a_o}{z}$

Where n is the orbit number and z is the atomic number of the atom. While $a_o = 0.0529 \times 10^{-9} nm$

Results:

When the angle θ_i is changing from 0 to 360 degrees and the $\Delta\theta = 0.1$ degree; we find the orbit shape of H atom and He atom by EXCEL; as the following:

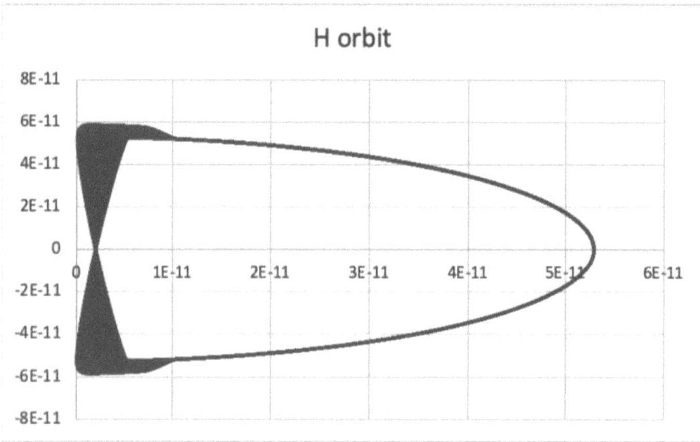

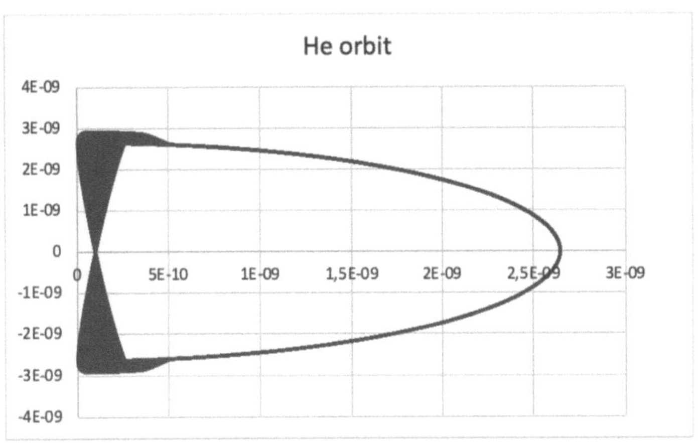

Now, to calculate the velocity, we find its value from the following steps;

$$t_i = \frac{x_i \sin \theta_i}{v_{y,i} \cos \theta_i} = \frac{x_i \tan \theta}{v_{y,i}} \quad \text{and} \quad t_i = \frac{y_i \cos \theta_i}{v_{x,i} \sin \theta_i} = \frac{y_i}{v_{x,i} \tan \theta}$$

$$\frac{x_i \tan \theta}{v_{y,i}} = \frac{y_i}{v_{x,i} \tan \theta}$$

$$\frac{v_{x,i}}{v_{y,i}} = \frac{y_i}{x_i \tan(\theta)^2}$$

$$v_{x,i} = v_{y,i} \frac{y_i}{x_i \tan(\theta)^2}$$

$$v^2 = v_{x,i}^2 + v_{y,i}^2$$

$$v^2 = \left(v_{y,i} \frac{y_i}{x_i \tan(\theta)^2}\right)^2 + v_{y,i}^2$$

$$v^2 t_i^2 = t_i^2 \left(v_{y,i} \frac{y_i}{x_i \tan(\theta)^2}\right)^2 + v_{y,i}^2 t_i^2$$

$$v^2 t_i^2 = y_i^2 \left(1 + \frac{y_i^2}{x_i^2 + \tan(\theta)^4}\right)$$

$$v^2 = \frac{y_i^2}{t_i^2} \left(1 + \frac{y_i^2}{x_i^2 + \tan(\theta)^4}\right)$$

$$v^2 = v_{y,i}^2 \left(1 + \frac{y_i^2}{x_i^2 + \tan(\theta)^4}\right)$$

$$v_{x,i}^2 = v_{y,i}^2 \left(1 + \frac{y_i^2}{x_i^2 + \tan(\theta)^4}\right) - v_{y,i}^2$$

$$v_{x,i}^2 = v_{y,i}^2 \left(\frac{y_i^2}{x_i^2 + \tan(\theta)^4}\right)$$

$$v_{x,i} = \frac{v_{y,i} y_i}{x_i \tan(\theta)^2}$$

$$v_{x,i} t_i = \frac{v_{y,i} y_i}{x_i \tan(\theta)^2} t_i$$

$$y_i t_i = x_i \tan\theta \rightarrow v_{x,i} t_i = \frac{y_i v_{y,i}}{v_{y,i} t_i \tan\theta}$$

$$v_{x,i} = \frac{y_i}{t_i^2 \tan\theta}$$

$$v_{x,i} t_i = \frac{y_i}{\tan\theta}$$

$$v_{x,i} = \frac{y_i}{t_i^2 \tan\theta} = \frac{y_i v_{x,i}^2 \tan(\theta)^2}{y_i^2 \tan\theta}$$

$$v_{x,i} = \frac{y_i^2}{\tan\theta}$$

$$v_{x,i} = \frac{y_i^2}{\tan\theta}$$

$$v_{y,i} = \frac{v_{x,i} x_i \tan(\theta)^2}{y_i} = \frac{y_i^2}{\tan\theta} \frac{x_i \tan(\theta)^2}{y_i}$$

$$v_{y,i} = y_i x_i \tan\theta$$

So, $v_{x,i} = \frac{y_i^2}{\tan\theta}$, $v_{y,i} = y_i x_i \tan\theta$, and $v = \sqrt{v_{x,i}^2 + v_{y,i}^2}$

The following graph shows the speed versus positions and velocities respectively of the electron in Hydrogen atom:

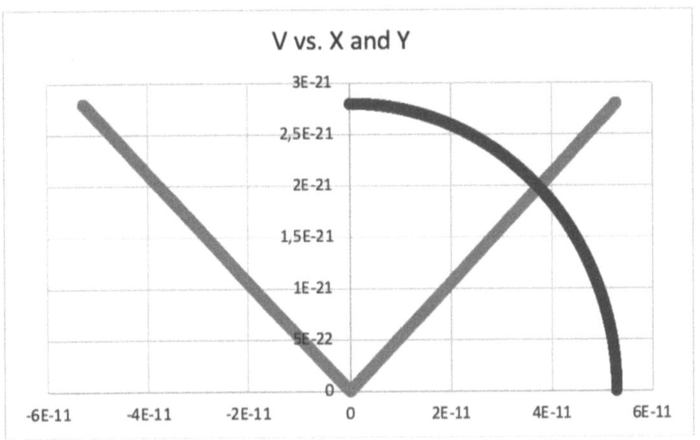

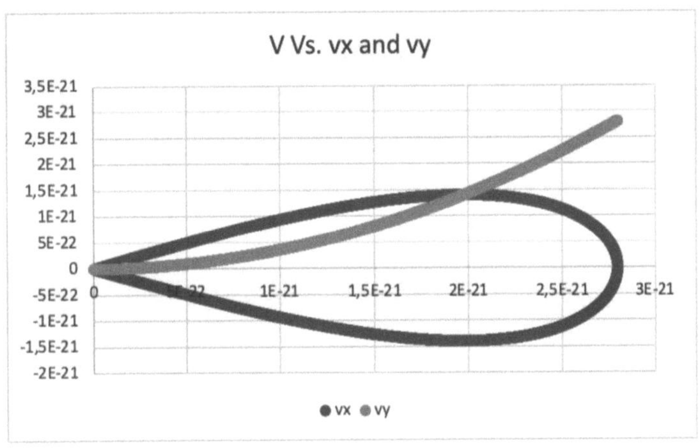

Comparing With Others:

By university of Tokyo; [10]

Superconductivity is a phenomenon in which electric circuit loses its resistance and becomes extremely efficient under certain conditions. Okazaki and his team used the method of ultralow-temperature and high-energy resolution laser-based photoemission spectroscopy to observe the way electrons behaved during a material's transition from The Bardeen-Cooper-Shrieffer (BCS) to Bose-Einstein condensate (BEC).

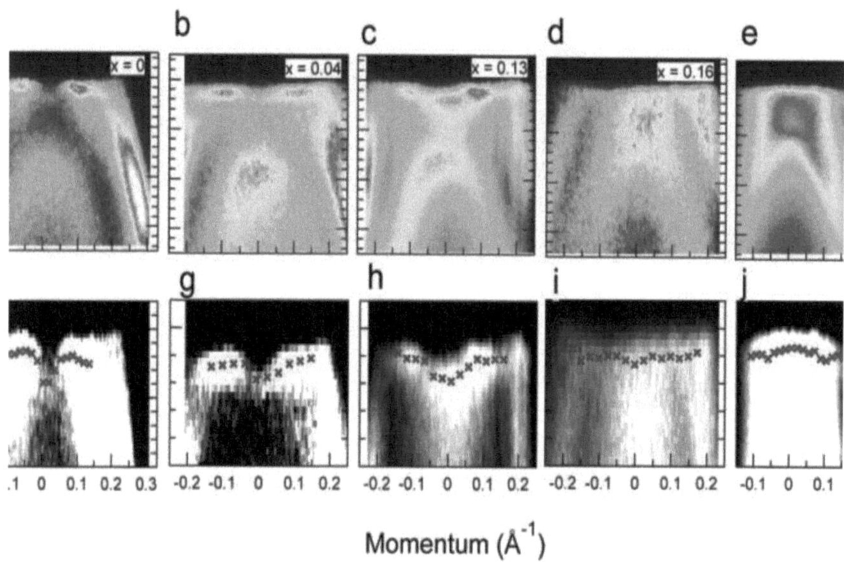

Polarized light images show researchers how electrons, represented by red crosses, in their test samples behave under different circumstances. Credit: © 2020 Okazaki et al.

Occult Chemistry
Clairvoyant Observations on the Chemical Elements
By

Annie Besant P.T.S. and
Charles W. Leadbeaterr
1919: [11]

"I remember the occasion vividly, Mr. Leadbeater was then saying at my house, and his clairvoyant faculties were frequently exercised for the benefit of myself, my wife and the theosophical friends around us. I had discovered that these faculties, exercised in the appropriate direction. Were ultramicroscopic in their power"

"It occurred to me once to ask. Mr. Leadbeater if he thought he could actually see a molecule of physical matter"

"he was willing to try, and I suggested a molecule of gold as one which he might try to observe".

"He made the appropriate effort, and emerged from it saying the molecule in question was far too elaborate a structure to be described it evidently consisted of an enormous number of some smaller atoms, quite too many to count; quite too complicated in there arrangement to be comprehended".

"It struck me at once that this might be due to the fact that the gold was heavy metal be more successful if directed to a body of low atomic weight, so I suggested an atom of hydrogen as possible more manageable."

"Mr. Leadbeater accepted the suggestion and tried again. This time he found the atom of hydrogen to be for simpler than the other, so that the minor atoms constituting the hydrogen atom were countable".

"They were arranged on a definite plan, which will be rendered intelligible by diagrams later on, and were eighteen in number".

"We little realized at the moment the enormous significance of this discovery, made in the year of 1895, long before the discovery of radium enabled physicists of the ordinary type to improve their acquaintance with "electron"."

"Whatever name is given to that minute body it is recognized now by ordinary science as well as by occult observation, as the fundamental unit of physical matter".

" To that extinct ordinary science has overtaken the occult research I am dealing with, it is perfectly certain, the ordinary physicists must follow him at no distant date."

"The research once started in the way I have described was seen to be intensely interesting".

"Mrs. Besant almost immediately co-operated with Mr. Leadbeater in its farther progress. Encouraged by the success with hydrogen, the two important gases, oxygen and nitrogen, were examined. They proved to be rather more difficult to deal with than hydrogen but were manageable".

" Oxygen was found to consist of 290 minor atoms and nitrogen of 261."

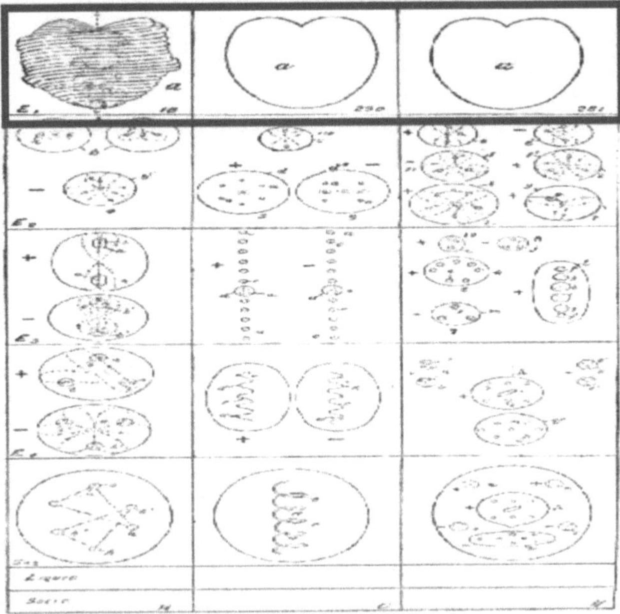

Figure: 34

The first chemical atom selected for this examination was an atom of hydrogen (H). In looking carefully at it, it was seen to consist of six small bodies contained in an egg-like form. It rotated with great rapidity on its own axis, vibrating at the same time, the internal bodies performed similar gyrations. The whole atoms spin and quivers and has to be steadied before exact observation is possible. The six little bodies are arranged in two sets of three, forming two triangles that are not interchangeable, but are related to each other as object and image.

The wall of the limiting spheroid in which the bodies are enclosed being composed of the third kind drops away when the gaseous atom is raised to the next level, and the six bodies are set free.

They at once re-arrange themselves in two triangles, each enclosed by a limiting sphere, the two marked b in the diagram unite with one of those marked b' to form a body which shows a positive character, the remaining three forming a second body negative in type. These forms the hydrogen particles.

Turning to the force side of the atom and its combinations, we observe that force pours in the heart-shaped depression at the top of the atom.

Speaking generally, positive bodies are marked by their contained atoms setting their points towards each other and the center of their combination and repelling each other outwards. Negative bodies are marked by the heart-shaped depressions being turned inwards, and by a tendency to move towards each other instead of away.

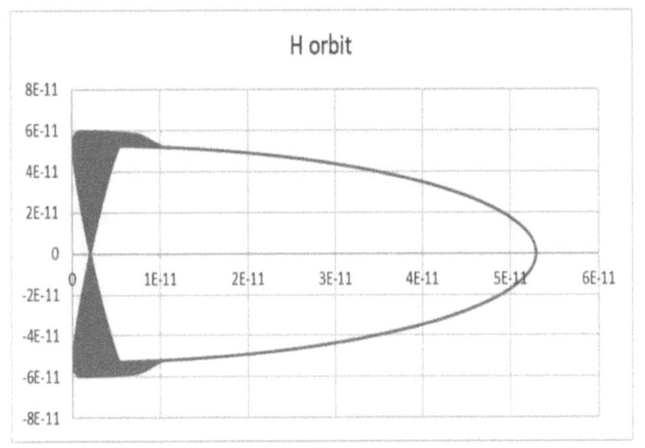

Occult Chemistry

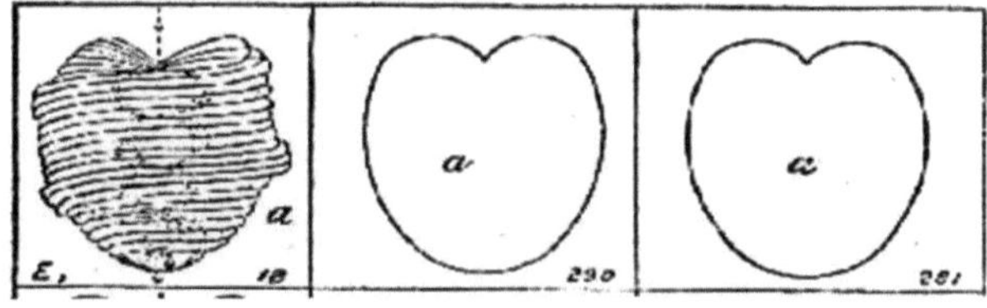

String Theory: From Newton to Einstein and Beyond:[12]

To understand the ideas and aims of string theory, it's useful to look back and see how physics has developed from Newton's time to the present day. One crucial idea that has driven physics since Newton's time is that of unification: the attempt to explain seemingly different phenomena by a single overarching concept. Perhaps the first example of this came from Newton himself, who in his 1687 work Principia Mathematical explained that the motion of the planets in the solar system, the motion of the Moon around the Earth, and the force that holds us to the Earth are all part of the same thing: the force of gravity.

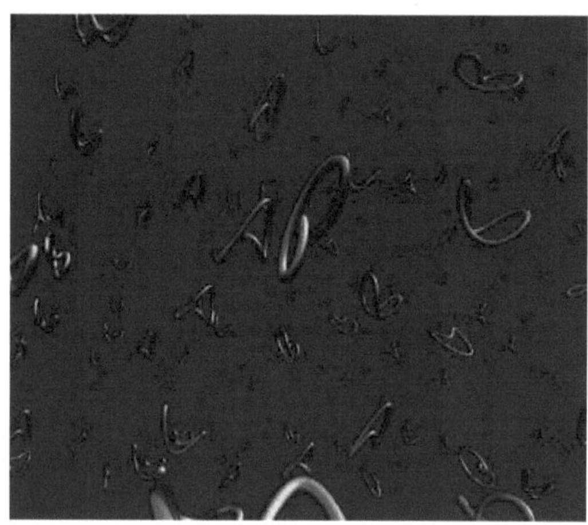

We take this for granted today, but pre-Newton the connection between a falling apple and the orbit of the Moon would have been far from obvious and quite amazing.

The next key unifying discovery was made around 180 years after Newton by the Scottish mathematician James Clerk Maxwell. Maxwell showed that electrostatics and magnetism, by no means similar phenomena at first sight, are just different aspects of a single thing called electromagnetism. In the process Maxwell discovered electromagnetic waves, which are in fact light — Maxwell had inadvertently explained a further seemingly different aspect of nature.

Another two hundred years on, in 1984, the Pakistani Abdus Salam and the American Steven Weinberg showed that the electromagnetic force and the weak nuclear force, which causes radioactive decay, are both just different aspects of a single force called the electroweak force.

This leaves us with three fundamental forces of nature: gravity, the electroweak force and the strong nuclear force which holds protons together.

Unifying matter

That deals with the forces, but what about matter? Many ancient belief systems have postulated that matter — and reality itself — is made from a finite number of elements. Modern physics confirms this idea. Experiments performed with the particle accelerator at CERN in Geneva have shown that there are just twelve basic building blocks of matter. These are known as the elementary particles. Everything we've ever seen in any experiment, here or in distant stars, is made of just these twelve elementary particles. (See Plus article The physics of elementary particles to find our more.)

All this is truly impressive: the entire Universe, its matter and dynamics explained by just three forces and twelve elementary objects. It's good, but we'd like to do better, and this is where string theory first enters: it is an attempt to unify further. To understand this, we have to tell another story.

Quantum gravity

There have been two great breakthroughs in the 20th century physics. Perhaps the most famous is Einstein's theory of general relativity. The other equally impressive theory is quantum mechanics.

General relativity is itself a unification. Einstein realised that space and time are just different aspects of a single object he called spacetime. Massive bodies like planets can warp and distort spacetime, and gravity, which we experience as an attractive force, is in fact a consequence of this warping. Just as a pool ball placed on a trampoline will create a dip that a nearby marble will roll into, so does a massive body like a planet distort space, causing nearby objects to be attracted to it.

The predictions made by general relativity are remarkably accurate. In fact, most of us will have inadvertently taken part in an experiment that tests general relativity: if it were false, then global positioning systems would be wrong by about 50 meters per day. The fact that GPSs work to within five meters in ten years shows just how accurate general relativity is.

The other great breakthrough of the 20th century was quantum mechanics. One of the key ideas here is that the smaller the scale at which you look at the world, the more random things become. *Heisenberg's uncertainty principle* is perhaps the most famous example of this. The principle states that when you consider a moving particle, for example an electron orbiting the nucleus of an atom, you can never ever measure both its position and its momentum as accurately as you like. Looking at space at a minuscule scale may allow you to measure position with a lot of accuracy, but there won't be much you can say about momentum. This isn't because your measuring instruments are imprecise. There simply isn't a "true" value of momentum, but a whole range of values that the momentum can take, each with a certain probability. In short, there is randomness. This randomness appears when we look at particles at a small enough scale. The smaller one looks, the more random things become!

The idea that randomness is part of the very fabric of nature was revolutionary: it had previously been taken for granted that the laws of physics didn't depend on the size of things. But in quantum mechanics they do. The scale of things does matter, and the smaller the scale at which you look at nature, the more different from our everyday view of the world it becomes randomness dominates the small scale world.

Again, this theory has performed very well in experiments. Technological gadgets that have emerged from quantum theory include the laser and the microchip that populate every computer, mobile phone and MP3 player.

But what happens if we combine quantum mechanics and relativity? According to relativity, spacetime is something that can stretch and bend. Quantum mechanics says that on small scales things get random. Putting these two ideas together implies that on very small scales spacetime itself becomes random, pulling and stretching, until it eventually pulls itself apart.

Evidently, since spacetime is here and this hasn't happened, there must be something wrong with combining relativity and quantum mechanics. But what? Both these theories are well-tested and believed to be true.

Perhaps we have made a hidden assumption? It turns out that indeed we have. The assumption is that it's possible to consider smaller and smaller distances and get to the point where spacetime pulls itself apart. What has rested in the back of our minds is that the basic indivisible building blocks of nature are point-like — but this may not necessarily be true

Strings to the rescue:

This is where string theory comes to the rescue. It suggests that there is a smallest scale at which we can look at the world: we can go that small but no smaller. String theory asserts that the fundamental building blocks of nature are not like points, but like strings: they have extension, in other words they have length. And that length dictates the smallest scale at which we can see the world.

What possible advantage could this have? The answer is that strings can vibrate. In fact they can vibrate in an infinite number of different ways. This is a natural idea in music. We don't think that every single sound in a piece of music is produced by a different instrument; we know that a rich and varied set of sounds can be produced by even just a single violin. String theory is based on the same idea. The different particles and forces are just the fundamental strings vibrating in a multitude of different ways.

The mathematics behind string theory is long and complicated, but it has been worked out in detail. But has anyone ever seen such strings? The honest answer is "no". The current estimate of the size of these strings is about 10^{-34}m, far smaller than we can see today, even at CERN. Still, string theory is so far the only known way to combine gravity and quantum mechanics, and its mathematical elegance is for many scientists' sufficient reason to keep pursuing it.

The theory's predictions:

If string theory is indeed an accurate model of spacetime, then what else does it tell us about the world?

One of its more startling and most significant predictions is that spacetime is not four, but ten-dimensional. It is only in ten dimensions of spacetime that string theory works. So where are those six extra dimensions? The idea of hidden dimensions was in fact put forward many years before the advent of string theory by the German Theodor Kaluza and the Swede Oskar Klein.

Shortly after Einstein described the bending of space in general relativity, Kaluza and Klein considered what would happen if a spatial dimension would bend round and rejoin itself to form a circle. The size of that circle could be very small, perhaps so small that it couldn't be seen. Those dimensions could then be hidden from view. Kaluza and Klein did show that in spite of this, these dimensions could still have an effect on the world we perceive. Electromagnetism becomes a consequence of the hidden circle with motion in the hidden dimension being electric charge. Hidden dimensions are possible and they in fact can give rise to forces in the dimensions that we can see.

String theory has embraced the Kaluza-Klein idea and currently various experiments are being devised to try and observe the hidden dimensions. One hope is that the extra dimensions may have left an imprint on the *cosmic microwave background*, the left-over radiation from the Big Bang, and that a detailed study of this radiation may reveal them. Other experiments are more direct. The force of gravity depends crucially on the number of dimensions, so by studying gravitational forces at short distances one can hope to detect deviations from Newton's law and again see the presence of extra dimensions.

Mathematics and physics have always influenced each other, with new mathematics being invented to describe nature, and old mathematics turning out to lend perfect descriptions for newly-discovered physical phenomena. String theory is no different and many mathematicians work on ideas inspired by it. These include the possible geometries of the hidden dimensions, the basic ideas of geometry when there is a minimum distance, the ways in which strings can split and come together, and the question of how we can relate strings to the particles in the world that we see.

String theory gives us an exciting vision of nature as miniscule bits of vibrating string in a space with hidden curled-up dimensions. All the implications of these ideas are yet to be understood. String theory is an active area of research with hundreds of people working to see how the theory fits together and produces the world we see around us.

References:

[1] En.wikipedia.org. 2020. *Atomic Theory*. [online] Available at: <https://en.wikipedia.org/wiki/Atomic_theory> [Accessed 10 December 2020].

[2] Encyclopedia Britannica. 2020. *Thomson Atomic Model | Description & Image*. [online] Available at: <https://www.britannica.com/science/Thomson-atomic-model> [Accessed 10 December 2020].

[3] En.wikipedia.org. 2020. *Atomic Theory*. [online] Available at: <https://en.wikipedia.org/wiki/Atomic_theory> [Accessed 10 December 2020].

[4] En.wikipedia.org. 2020. *Atomic Theory*. [online] Available at: <https://en.wikipedia.org/wiki/Atomic_theory> [Accessed 10 December 2020].

[5] Pbs.org. 2020. *A Science Odyssey: People And Discoveries: Erwin Schrodinger*. [online] Available at: <http://www.pbs.org/wgbh/aso/databank/entries/bpschr.html> [Accessed 10 December 2020].

[6] Ch301.cm.utexas.edu. 2020. *Orbital Shapes*. [online] Available at: <https://ch301.cm.utexas.edu/section2.php?target=atomic/H-atom/orbital-shape.html> [Accessed 10 December 2020].

[7] Chemguide.co.uk. 2020. *Atomic Orbitals*. [online] Available at: <https://www.chemguide.co.uk/atoms/properties/atomorbs.html> [Accessed 10 December 2020].

[8] En.wikipedia.org. 2020. *Uncertainty Principle*. [online] Available at: <https://en.wikipedia.org/wiki/Uncertainty_principle> [Accessed 10 December 2020].

[9] En.wikipedia.org. 2020. *Atomic Orbital*. [online] Available at: <https://en.wikipedia.org/wiki/Atomic_orbital> [Accessed 9 November 2020].

[10] Phys.org. 2020. *Researchers Demonstrate A Superconductor Previously Thought Impossible*. [online] Available at:

<https://phys.org/news/2020-11-superconductor-previously-thought-impossible.html> [Accessed 9 November 2020].

[11] Leadbeater, C., n.d. *Occult Chemistry Clairvoyant Observations On The Chemical Elements*. [Place of publication not identified]: Project Gutenberg.

[12]"Atomic Theory Timeline." Timetoast, www.timetoast.com/timelines/atomic-theory-timeline-46a0ffa4-007b-40b6-825c-1f9dec8df530. Accessed 11 Dec. 2020.

Carpi, Anthony. "Early Ideas about Matter: From Democritus to Dalton." Visionlearning, Visionlearning, Inc., 17 Mar. 2003, www.visionlearning.com/en/library/Chemistry/1/Early-Ideas-about-Matter/49.

"Joseph Proust." Wikipedia, 27 Nov. 2020, en.wikipedia.org/wiki/Joseph_Proust. Accessed 11 Dec. 2020.

"Leucippus." Wikipedia, 6 Dec. 2020, en.wikipedia.org/wiki/Leucippus. Accessed 11 Dec. 2020.

"Leucippus"---. Wikipedia, 6 Dec. 2020, en.wikipedia.org/wiki/Leucippus.

"Matter." Abyss.Uoregon.Edu, abyss.uoregon.edu/~js/ast121/lectures/lec07.html. Accessed 11 Dec. 2020.

PITTORESQUE, LA FRANCE. "8 Mai 1794 : Exécution Du Chimiste Antoine-Laurent Lavoisier." La France Pittoresque. Histoire de France, Patrimoine, Tourisme, Gastronomie, 1 Jan. 1970, www.france-pittoresque.com/spip.php?article7750.

"String Theory: From Newton to Einstein and Beyond." Plus.Maths.org, Dec. 2007, plus.maths.org/content/string-theory-newton-einstein-and-beyond.

Walia, Ruhani. "The Socratic Method." Medium, 29 Oct. 2019, ruhani-walia.medium.com/the-socratic-method-661ef6d14bf7. Accessed 11 Dec. 2020.

Figure References:

- https://www.compoundchem.com/2016/10/13/atomicmodels
- https://www.sutori.com/item/1803-dalton-s-atomic-theory-john-dalton-publishes-his-atomic-theory-which-states
- https://www.sciencephoto.com/media/224559/view
- https://www.france-pittoresque.com/spip.php?article7750
- https://www.visionlearning.com/en/library/Chemistry/1/Early-Ideas-about-Matter/49
- https://en.wikipedia.org/wiki/Joseph_Proust
- http://abyss.uoregon.edu/~js/ast121/lectures/lec07.html
- https://ruhani-walia.medium.com/the-socratic-method-661ef6d14bf7
- https://www.timetoast.com/timelines/atomic-theory-timeline-46a0ffa4-007b-40b6-825c-1f9dec8df530
- https://en.wikipedia.org/wiki/Leucippus
- https://www.aaas.org/news/lucretius-science-verse
- https://en.wikipedia.org/wiki/Epicurus
- http://www.bbc.co.uk/ahistoryoftheworld/objects/qBh00JqDTNq6YQRX-mz4Tg
- http://ieros.info/Site/Lives_and_Opinions_of_Eminent_Philosophers_%28Vol_II%29/Lives_and_Opinions_of_Eminent_Philosophers_%28Vol_II%29.html
- https://www.alamy.com/stock-photo-pierre-gassendi-1592-1655-was-a-french-philosopher-priest-scientist-104000976.html
- https://www.google.com/url?sa=i&url=https%3A%2F%2Fwww.youtube.com%2Fwatch%3Fv%3DUHZO-ipqtcc&psig=AOvVaw1DvTEg0d4j3txf6Enyxqbp&ust=1607163973750000&source=images&cd=vfe&ved=0CAIQjRxqFwoTCLDv6ciOtO0CFQAAAAAdAAAAABAE
- https://www.google.com/url?sa=i&url=https%3A%2F%2Fwww.azquotes.com%2Fauthor%2F28103-Robert_Boyle&psig=AOvVaw3UqsetszzUq7GKxU3eOl4M&ust=1607163818172000&source=images&cd=vfe&ved=0CAIQjRxqFwoTCOjCq4OOtO0CFQAAAAAdAAAAABAK
- http://koyre.ehess.fr/docannexe/file/905/bigg_evident_atoms_shps.pdf
- https://es-la.facebook.com/BFScientificLife/photos/jean-baptiste-perrin-was-a-french-physicist-who-in-his-studies-of-the-brownian-m/1030249123852427

- https://en.wikipedia.org/wiki/Plum_pudding_model
- https://micro.magnet.fsu.edu/optics/timeline/people/rutherford.html
- https://micro.magnet.fsu.edu/optics/timeline/people/rutherford.html
- https://www.pinterest.com/pin/188799409357984939/
- https://en.wikipedia.org/wiki/File:Rutherford_gold_foil_experiment_results.svg
- https://www.cambridge.org/core/journals/microscopy-today/article/pioneers-in-optics-niels-henrik-david-bohr/4FB44B35EC2DC51D027571C03E5C4D4F
- https://en.wikipedia.org/wiki/Bohr_model
- https://en.wikipedia.org/wiki/Arnold_Sommerfeld
- https://en.wikipedia.org/wiki/Photoelectric_effect
- https://en.wikipedia.org/wiki/Erwin_Schr%C3%B6dinger
- https://www.chemicool.com/definition/schrodinger_equation.html
- http://abyss.uoregon.edu/~js/21st_century_science/lectures/lec14.html